HEATH SCIENCE

HEATH SCIENCE

Authors

James P. Barufaldi
Professor of Science Education
University of Texas, Austin

George T. Ladd
Professor of Education
Boston College

Alice Johnson Moses
Elementary Science Teacher
University of Chicago Laboratory School

Consulting Authors

Herman and Nina Schneider

Content Consultants

Stephen S. Barshay
Assistant Professor of Chemistry
Rhode Island College
Providence, Rhode Island

Joan G. Creager
Professor of Science
Marymount College of Virginia
Arlington, Virginia

John A. Day
Professor of Physics and Meteorology Emeritus
Linfield College
McMinnville, Oregon

Victor J. Mayer
Professor of Science Education and Geology
Ohio State University
Columbus, Ohio

R. Robert Robbins
Piper Professor of Astronomy
University of Texas
Austin, Texas

Frederick J. Stare, M.D.
Professor of Nutrition Emeritus
Harvard University School of Public Health
Boston, Massachusetts

J. Kenneth Taylor
Chairman of the Biology Department
Westfield State College
Westfield, Massachusetts

Jearl Walker
Professor of Physics
Cleveland State University
Cleveland, Ohio

D.C. Heath and Company
Lexington, Massachusetts   Toronto

Project Editor: Paul P. Scopa

Assistant Editor: Roland E. Boucher

Book Designer: Eleanor Thayer

Production Coordinator: Donna Lee Porter

Reading Level Consultants: J & F
Milton Jacobson, *Founder*
Charlottesville, Virginia

Testing and Evaluation Consultant: Peter W. Airasian
Professor, Educational Testing and Evaluation
Boston College, Chestnut Hill, Massachusetts

Educational Consultants

Sister Marie Andre Guay, RSM
Catholic School Office
Providence, Rhode Island

Elaine S. Barrett
Science Project Teacher
Bellevue School District, Washington

Harriett Brookman
Elementary Teacher
Miami, Florida

Napoleon Bryant, Jr.
Professor of Education
Xavier University, Cincinnati, Ohio

Jerry Hayes
Elementary Science Supervisor
Chicago, Illinois

David Lawrence
District Chairperson Science
Hartford, Connecticut

Richard Merrill
Curriculum Specialist
Mt. Diablo Unified School District, Concord, California

Constance Tate
Coordinator, Office of Science
Baltimore, Maryland

Roger Van Bever
Supervisor of Elementary Science
Detroit, Michigan

Mary Ellen Wandel
Supervisory Instructional Specialist
Pittsburgh, Pennsylvania

Edwin P. White
School of Education
University of South Carolina, Spartanburg

Field Test Teachers

Grateful acknowledgment is given to the teachers and students who participated in field tests of Heath Science.

Phoenix, ARIZONA
Lookout Mountain School, Barbara Semmens
Acacia School, Rebecca Walker

Louisville, COLORADO
Louisville Elementary School, Dorothy Pecina, Cheryl Turvey

Boulder, COLORADO
Bear Creek Elementary School, Barbara Puzio, Beverly Vance, Virginia Holland

Merritt Island, FLORIDA
Audubon Elementary School, Sue Colley, Natalie Jerkins, Peggy Graham, Ann Ehren

Satellite Beach, FLORIDA
Surfside Elementary School, Crystal Otto, Judith Cordrey, Sue Wisler

Atlanta, GEORGIA
E. Rivers School, Lillie Thompson, Virginia Robb, Cecelia J. Thomas, Sandra L. Black

Lexington, MASSACHUSETTS
Bowman Elementary School, Maureen Sullivan
Maria Hastings School, Alice F. Baylies

Maynard, MASSACHUSETTS
Coolidge School, Ellen Holway, Elizabeth Niland
Green Meadow School, Daria Benham, Pamela Tiramani

Newton, MASSACHUSETTS
Oak Hill School, Francis R. Stec, Dorothy Mims

Brooklyn, NEW YORK
Blessed Sacrament School, Marilyn Robinson
St. Jerome School, Sr. Patricia Brennan, R.S.M.

Plainview, NEW YORK
Parkway School, Edna Publicker, Shirley Grant
Pasadena School, Muriel Phillips, Suzanne Freund

Norfolk, VIRGINIA
Mary Calcott Elementary School, Barbara Wright
Ocean Air School, Yolanda Hill
Engleside School, Naomi Bethea
Poplar Halls School, Christine Crouch

Cover and Frontispiece Photo: Shallow Reef Scene by Carl Roessler, Animals Animals
Photo illustration credits appear on page 249.

Published simultaneously in Canada.

Printed in the United States of America.

International Standard Book Number:
0-669-05607-3

TABLE OF CONTENTS

1 Seasons

Spring, summer, fall, and winter!
Every year has four parts or seasons.
What season is it now?
What season do you see in each picture?

It is fall.
You are back in school.
The air begins to get cold.
What do you wear?
What things do you do?

Chipmunk

Cecropia moth

Geese migrating

What do animals do in the fall?
Some animals store food.
Some get ready for a deep sleep.
What else do they do?
Some birds fly away.
But plants must stay.

In some places the leaves change color.
Then these leaves fall.
The leaves of some plants stay green.
Do the leaves near your home change color?

ACTIVITY

You can find many things in the fall.

What kinds of things can you find?

Winter comes after fall.
The days are short.
The air turns cold.
In some places snow falls.
You find warm things to wear.
What is winter like where you live?

I wonder, I wonder,
How animals know
To go to sleep
With the coming of snow.

Some animals rest in the winter.
Many plants rest in the winter, too.
They store food in their roots.
Other plants do not live in winter.
But they have made seeds.
The seeds will be ready to grow in the spring.

ACTIVITY

Get three plants of the same kind.
Put one in a cold place.
Keep the other plants in a warm place.
Cover one of them.
Wait one day.
Draw pictures of what you see.

Spring comes!
The days are not as short.
The air turns warm.
Where did the winter snow go?

Crocuses

Plants begin to grow.
Soon you can see flowers.
There are many things to do outside.

I wonder, I wonder,
How animals know
To come out in the spring
When other things grow.

Canada geese

Bison

Hereford cattle

The winter rest is over.
Plants begin to grow and change.
The air is full of sounds of animals again.
There are many baby animals.

ACTIVITY

What happens to seeds in the spring?

Get some seeds.

Plant them.

Water them.

Put both boxes in a cold place.

Wait a few days.

Do seeds grow in the cold?

Now put one box in a warm place.
Keep the other box in the cold place.
Wait a few more days.
What do you see?
Do seeds grow where it is warm?

Now it is summer!

What are the days like?

It is a time to swim and play.

What else can you do in the summer?

The plants grow tall.
There are many leaves and flowers.
Animals are busy doing many things.
The baby animals grow.

ACTIVITY

Each year has four seasons.
Which season do you like best?
Is it a warm or cold season?
Draw pictures about this season.

Show what you wear.
Show the things you like to do.
What things do you see?
What sounds do you hear?
What things do you smell and taste?
What things do you feel?
Show them in your pictures.

TO THINK ABOUT AND DO

SEASON HUNT

Can you find the names of the seasons?

Move the letters around.

rntiew

msmreu

llfa

gsinpr

NAME THE SEASON

In what season do animals do these things?

?   ?   ?   ?

SPRING!

Look at the picture.

What things tell you it is spring?

LOOK AT THE TREES

The trees tell a story.

Can you name each season?

HOW DO ANIMALS IN SOIL GET AIR?

In winter some animals sleep in the soil.

How can they get air?

Put some dry soil in a cup.

Then put some water in the cup.

Add more and more water.

What happens?

Sound

2

So many sounds!
Were you ever near a band like this?
What kinds of sounds did you hear?
How did they make you feel?

There are many different sounds.
Yet all sounds are alike in one way.
Let's find out.

ACTIVITY

Cut open a balloon.

Pull it over the top of a jar.

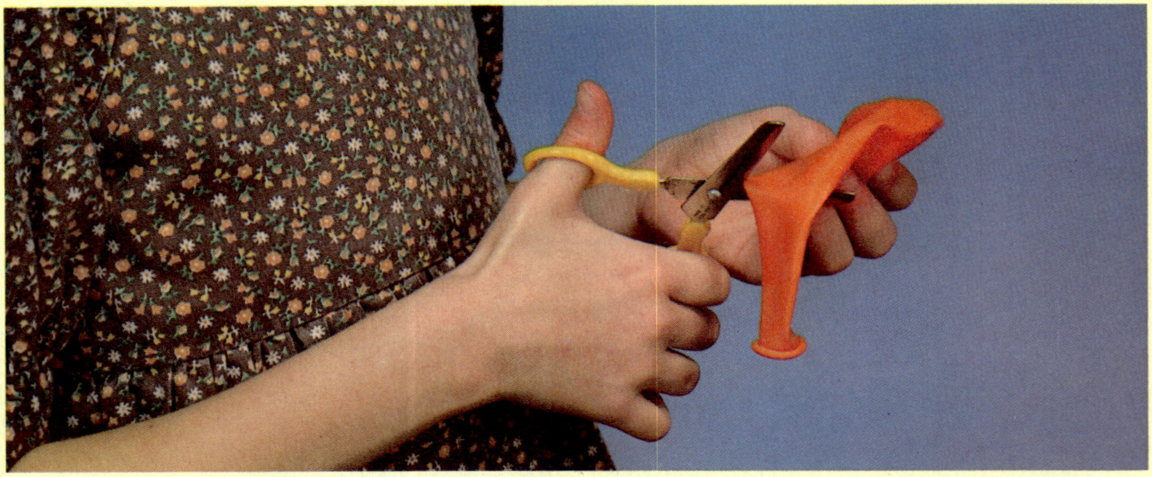

Now get a few seeds.
Put them on the balloon.
Tap the balloon.
What do you hear?
What happens to the seeds?
Now tap the balloon hard.
What do you hear?
How do the seeds move?
How does the balloon move?

ACTIVITY

Make one of these.
Make a sound with it.
Listen to the sound.

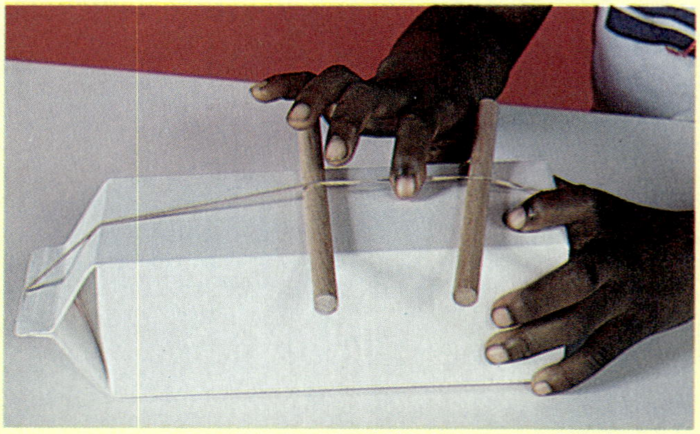

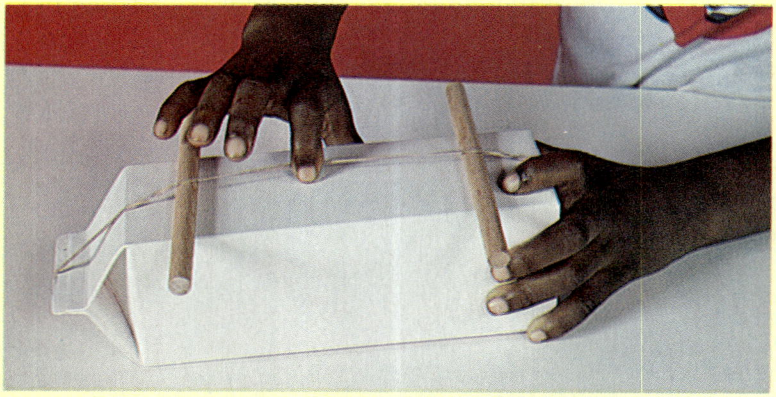

Now move one stick away from the other.
Make a sound again.
What do you hear?
Make a soft sound.
Make a high sound.
Make other sounds.
How did you make the sounds change?

ACTIVITY

Put one of these on a table.
Hold one end like this.
Push the other end.
Then let go.
Look at the way it moves.
Listen to the sound.

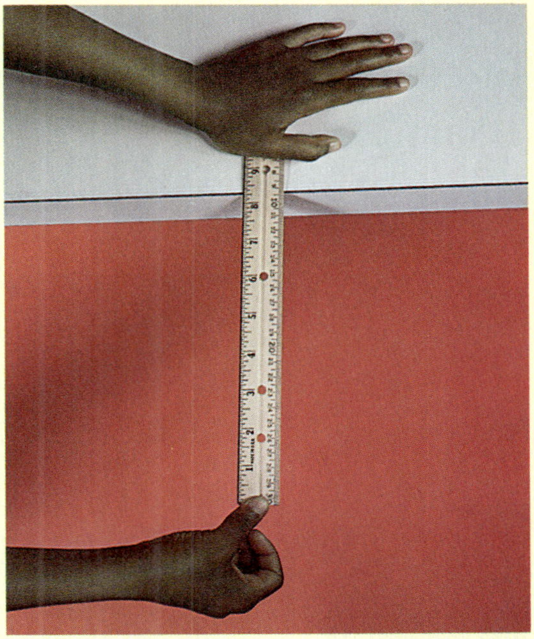

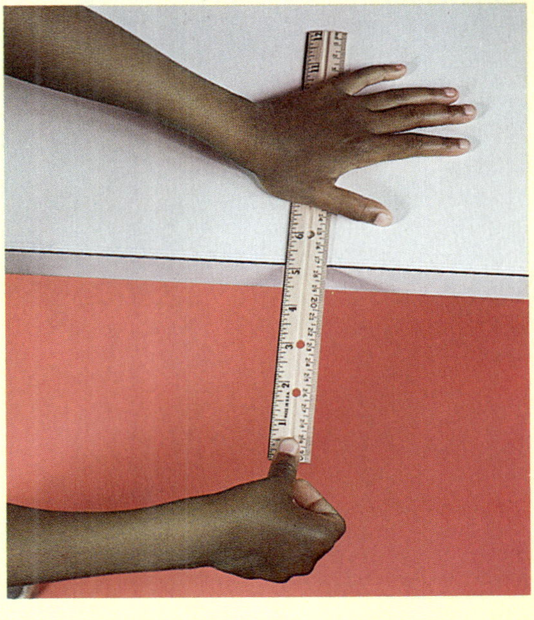

Now hold it like this.
Try it again.
Look at the way it moves now.
How did the sound change?

ACTIVITY

Make other things that make sounds.
Listen to the different sounds.
Then cover your eyes.
Guess what made the sound.

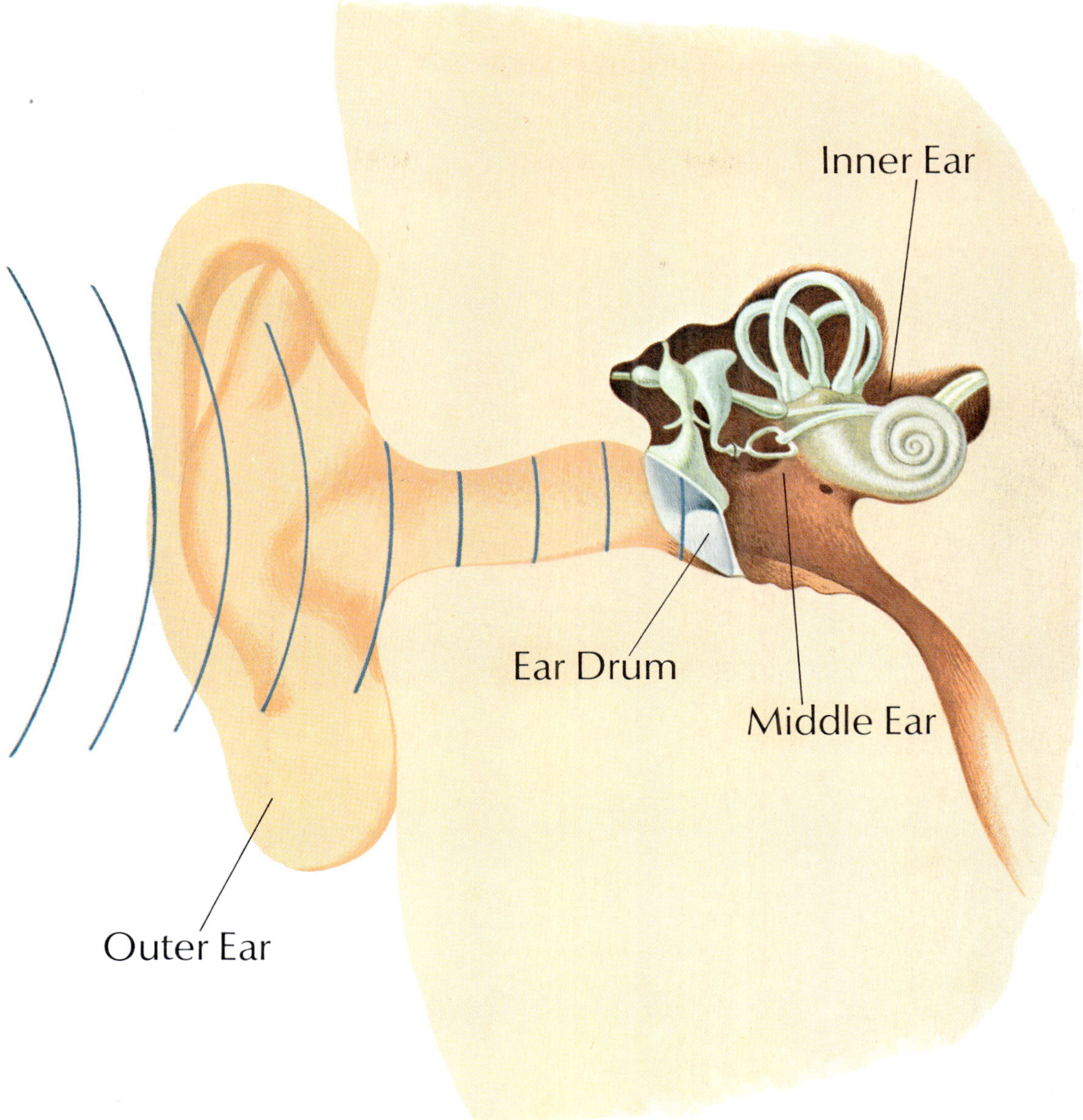

Many sounds come to our ears.
Our ears have many parts.
These parts help us hear.

ACTIVITY

Sometimes we want to hear better.
Hold a watch away from your ear.
How does it sound?

Now do this.
How is the sound different?

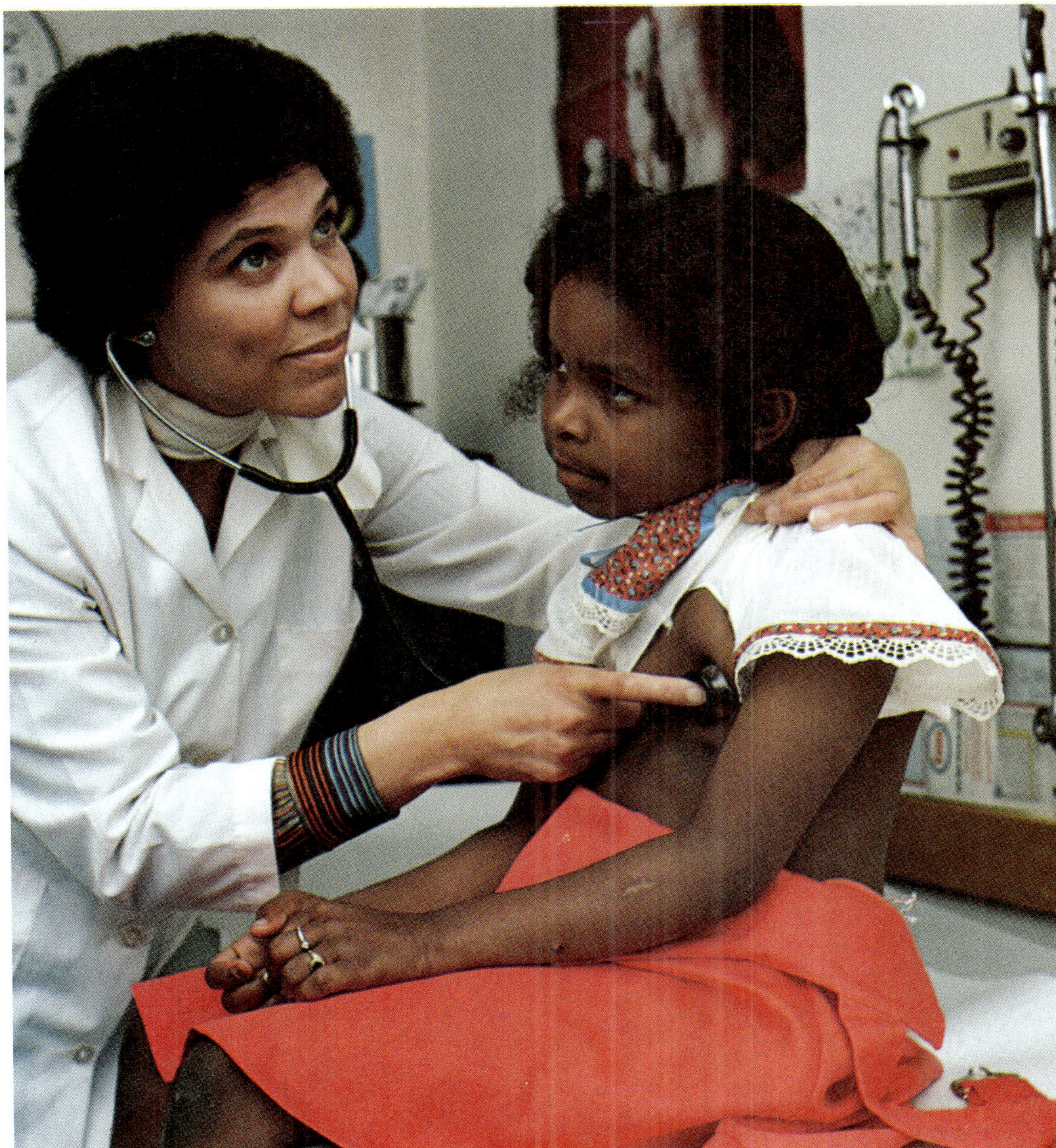

Have you seen one of these?
How does it help her hear?

We should take good care of our ears.
Keep them clean.
Never put things in your ears.
You might hurt them.

Very loud sounds can also hurt your ears.
What other places have very loud sounds?

What are these people doing?
How is the boy taking care of his ears?
How do you take care of your ears?

We can make different sounds.
We can hear different sounds.
We hear with our ears.

TO THINK ABOUT AND DO

LISTEN!

Make one of these.

Use two paper cups and some string.

Now talk to a friend.

How does the sound move?

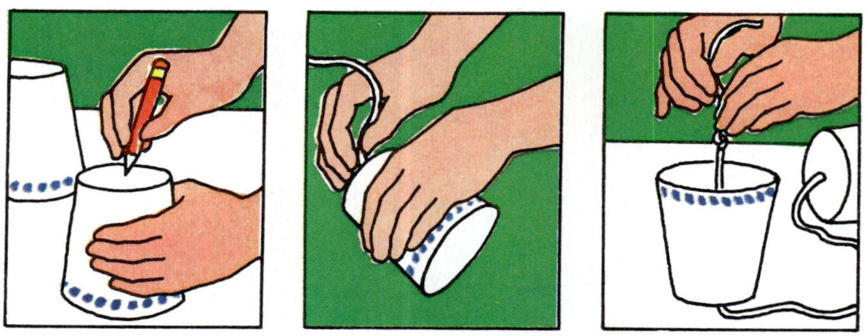

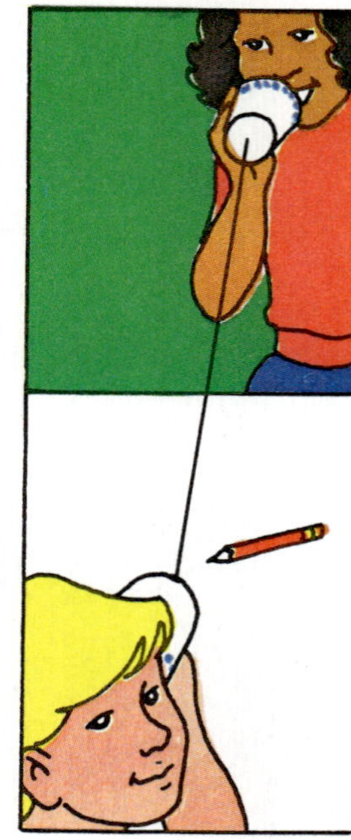

TRY THIS

Drop a small stone on the floor.

Then drop it in water.

Drop it on something soft.

How were the sounds different?

Which one made a loud sound?

Which one made a soft sound?

SOMETHING DIFFERENT

Tie a piece of string like this.

Hold the ends of the string to your ears.

Let it bump into things.

Listen to the sounds.

MANY WORDS TELL ABOUT SOUND

How many can you think of?

Here are some words to start.

bang

knock

roar

Look at your list of words for sounds.

Which sounds can be loud?

Which sounds can be soft?

Eating and Growing 3

Look at the foods!

How many of these foods can you name?

Did you ever eat any of them?

Which ones?

We eat many kinds of food.
We like some foods better than others.

This boy just ate one of these.
Which do you think it was?

ACTIVITY

Find a picture of one food you like best.

Show the picture.

Tell why you like the food.

We need to eat many kinds of food.
Foods help us in many ways.
Foods help us grow.
Foods help us work and play.
Some foods help to make strong teeth and bones.
We need many kinds of food each day.
What are they?

Food should be clean before we eat it.
Some things on the food can make us sick.
Our hands should be clean, too.
Why?

ACTIVITY

Put a drop of paint on your hand.

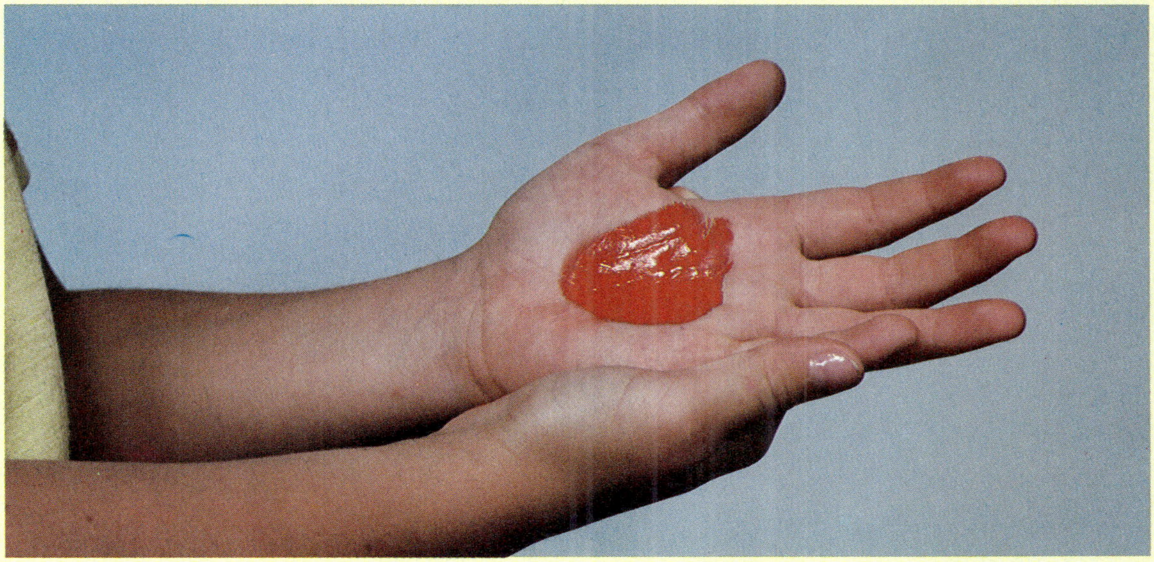

Look at your hand.
What do you see?

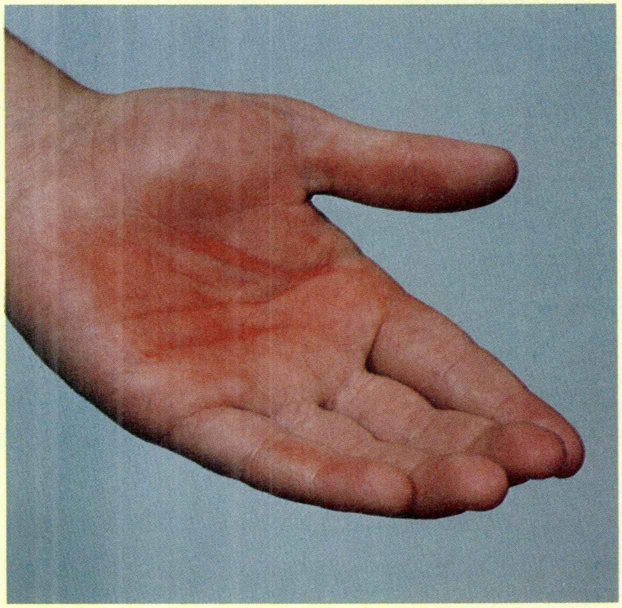

Wash your hand.
Did you wash it well?

Our teeth change food.
They chew food into small bits.
How does this help our body use food?

ACTIVITY

Get two of these.

Put one into a cup.
Break the other into small pieces.
Put it into another cup.
Which is like food after you eat it?
Now fill each cup with water.
Move the water around in the cups.
What happens?

ACTIVITY

How many teeth do you have?

How many have you lost?

How many new ones do you have?

Do you have these kinds of teeth?

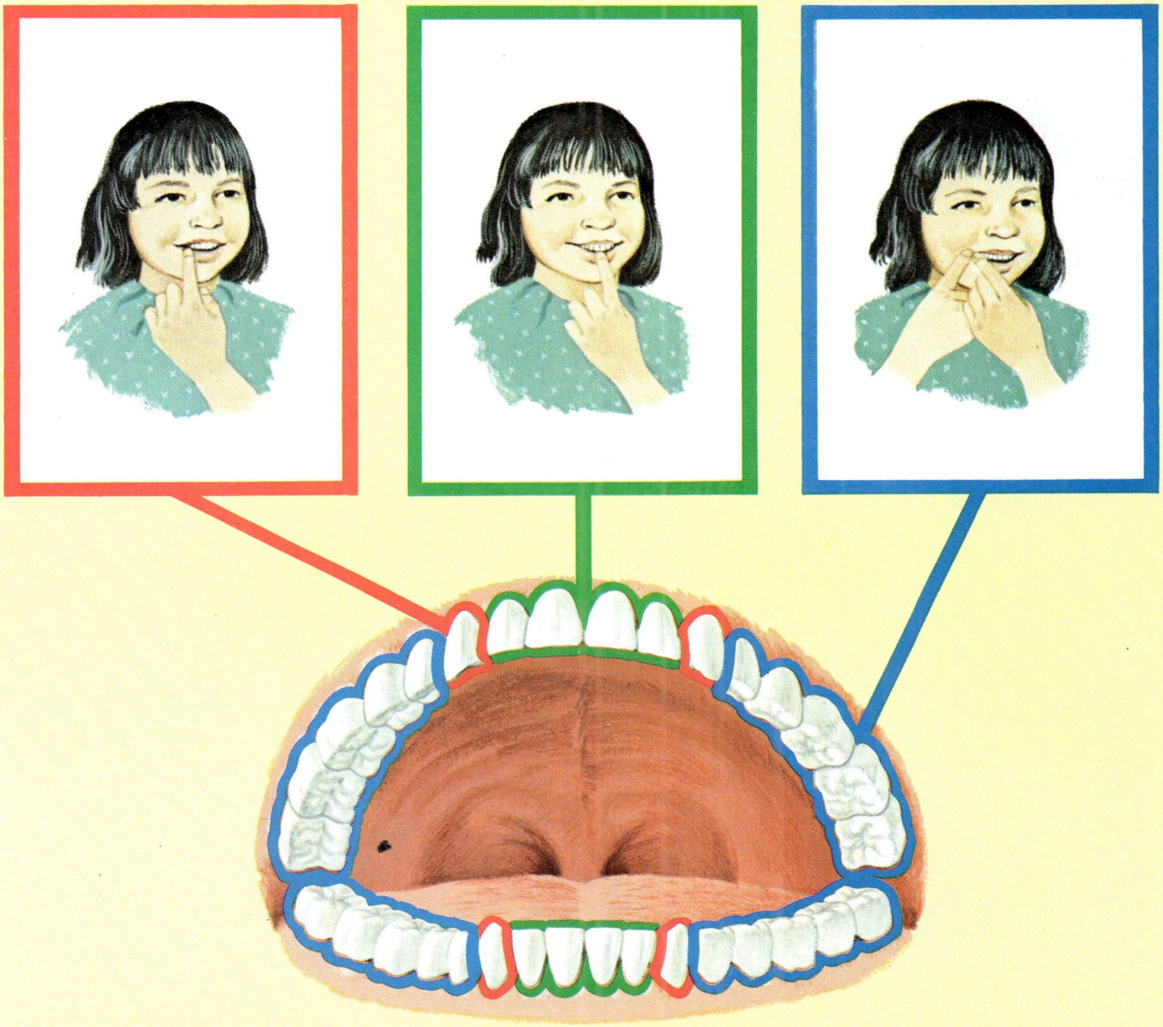

How are they different?
What does each kind do?

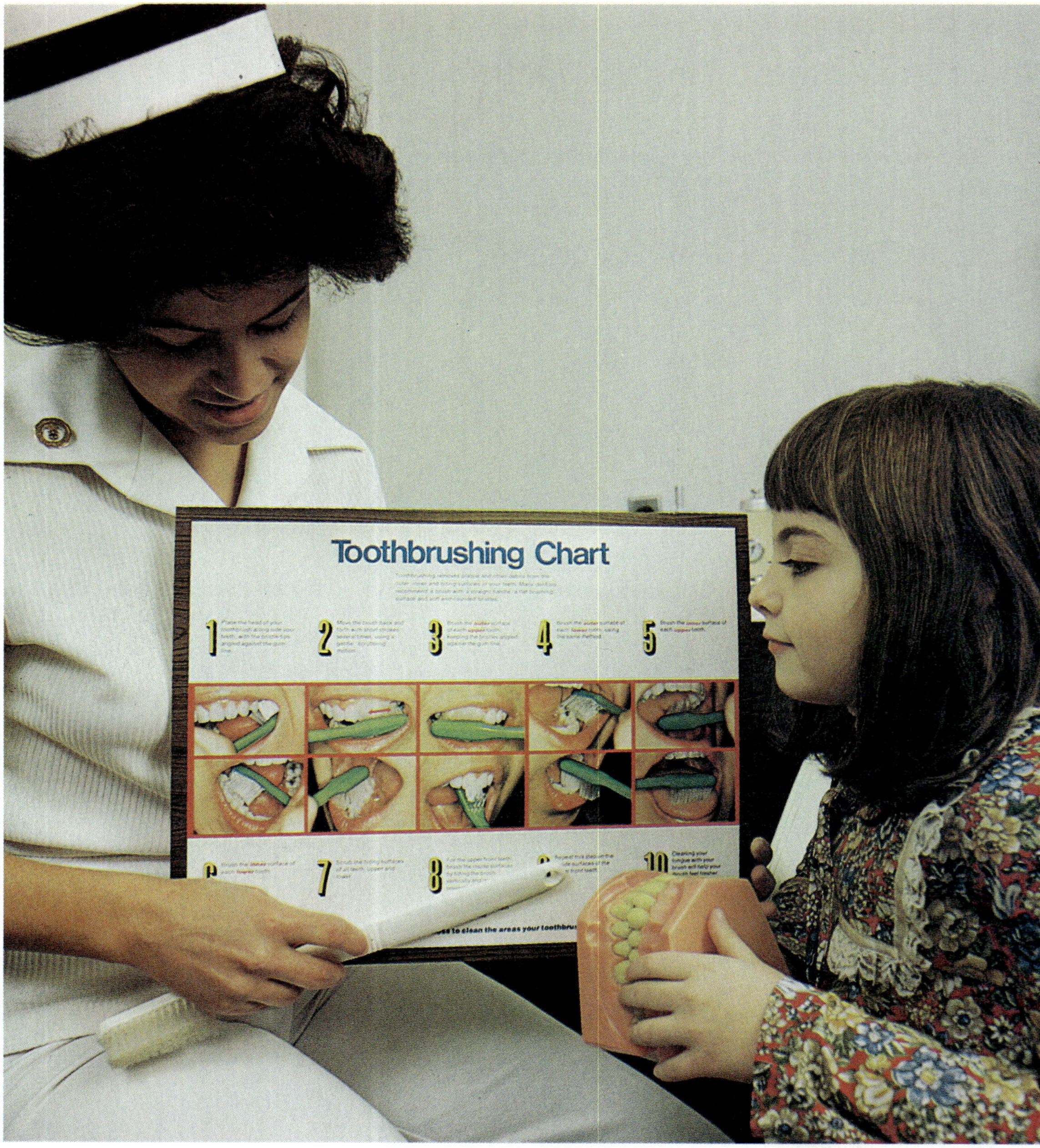

Do you take care of your teeth?
How do you clean them?

Some foods help make teeth hard and strong.
Eat some of these foods every day.

Once you were small.
Now you are bigger.

You can do many things.
You are growing and changing.

A baby needs care.
It can not do things on its own.
It is not ready to put on its own clothes.
It needs help to eat.

This girl is seven years old.
She is growing and changing.
She wants to be like her sister.
Her sister is ten years old.
Many changes take place as living things grow.

What is happening in each picture?
Animals change as they grow.
Plants change as they grow.

PLAN A DINNER

Look at some pictures of food.
What would you eat for dinner?
Make a picture of your dinner.

MAKE A PICTURE

Some foods help make strong teeth.
Draw pictures of the foods.

GROWING AND CHANGING

When you were a baby, you could do very little.
You could not feed yourself.
You could not walk or talk.
Now you can do many things on your own.
What are some of these things?

4 Light and Shadow

What a shadow!
Something big must have made it.
Or did something small make the shadow?
Try to find what made it.

You can make a shadow.
First you need a light.
Next you need an object.
Then you need a place to see the shadow.

ACTIVITY

Make a shadow.

What happens when you move?

Try to get away from your shadow.

Can you get away?

You can not have a shadow without light. Light comes from many things.

Where does most light come from in the day?
Where does it come from at night?

Many lights are made by people.
We use lights to help us see.
Lights help us know when to watch out.
Lights help us find things.
How have lights helped you?

Have you ever seen your shadow change like this?
What happens?

Light moves through some objects.
Other objects stop light.
Objects that stop light make shadows.

ACTIVITY

Try things like these.
What did you find out?

ACTIVITY

Draw a line around the shadow.

Draw a line around the object.

Are the shapes the same?

What do shadows show you about objects?

Let's play a shadow game.
Find something in the room.
Make a shadow with it.
Ask your friends what the object is.

Which objects made the shadows?

ACTIVITY

Can you make the shadow change?

Make the shadow big.

Then make the shadow small.

Make different shadows with the same object.

The sun shines on all these objects.
Find the tall objects.
Find the long shadows.

Find the short objects.
Find the short shadows.
Where do you think the sun is?

ACTIVITY

It is morning.
Where is the sun?
Look at another child's shadow.
Draw a line around it.
Put an X where the feet are.

Try it again after a while.
Have the child stand on the X.
Did the shadow change?
Did anything else change?
Where will the shadow be after school?
What will it look like?

Look at all the shadows!
Think about what you need to make them.
Wherever you go you can find shadows.
You can make shadows, too!

TO THINK ABOUT AND DO

PLAY A SHADOW GAME

Make a shadow rabbit.

Make other shadow animals.

Can your friends name them?

WHAT MADE THESE SHADOWS?

MAKE SOME SHADOWS

Hold something and make a shadow with it.
Make different shadows with the same thing.
Can you make different shadows of yourself?
Can you get away from your shadow?

LOOK AT THE SHADOWS

Which shadow does not belong?
Where does the light come from?

SHADOW FACES

Make a shadow picture of a friend's face.

Draw a line around the shadow.

Cut it out.

Make other shadow faces.

Can you tell who each is?

SHADOW WORDS

Shadows cover letters of these words.

What are the words?

sha ow

lig t

ob ect

s n

n ght

shap

5 Moving

All around you things are moving.
Some things run.
Some things roll.
There are even things that fly.

What other ways do things move?

Things that move change place.
You can see that things are moving.

How can you tell if some things have already moved?
Are they in the same place?

ACTIVITY

Put some things on a table.
Have other children look at them.
Now have them cover their eyes.
Move one of the things.
Have the children open their eyes.
Ask which one was moved.
How did they know?

The boy was here.
Then he was there.

Now he is over there!
His place is different each time.
How can you tell which way he is moving?

Things move in different ways.
Some things move in straight lines.
Some move like this.

Other things seem to move every which way.
Tell the way each thing is moving.

ACTIVITY

Roll a ball to another child.
Which way did the ball roll?
What did you see?
What did the children see?
Ask other children what they saw.
Does each one give the same answer?

Think about what each one sees.
Things look different from different places.

Hummingbird

Some things move fast.

Some things look as if they are not moving at all.

What is happening here?
Which one is moving fast?
Which is slow?
Which is in between?
How do you know?

Can you be sure what things are moving?
You and a friend are sitting in a bus.
Is your friend moving?
Are you moving?
What looks as if it is moving?
What is really moving?

What things seem to be moving here?
What things do not seem to be moving?
Now look again.

What things are moving here?
What things are not moving?
What are some other times when this could happen?

ACTIVITY

Find something here that moves.
Do not tell what it is.
Tell how the thing moves.
Ask another child to guess what it is.
Now have the child tell how another thing moves.
Can you guess what it is?

Many things move.

Cars and birds, trains and dogs.

All these things move.

And you move, too.

TO THINK ABOUT AND DO

DO NOT GET LOST

Begin at start.

Walk to the light and turn right.

Walk to the end of the street.

Then turn left.

Walk straight and cross two streets.

Where are you now?

WHAT MOVED?

Something moved.

What was it?

How can you tell?

MANY WAYS TO MOVE

Many words tell about moving.

How many can you think of?

Here are some.

fly   hop   swing

MOVING OR NOT MOVING?

How can the child in the car move and not move?

What other times do you move and not move?

6

Night and Day

Day, night, and day again!
Every day the sun warms us.
It gives us light.
But did you ever wonder where the sun is at night?

The sun rises and sets.
All day it gives us light.
The sun is hot and bright.

When the sun sets it gets dark outside.
Where is the sun at night?
Is it still shining?

Earth

This is a picture of the earth.
It shows how the earth looks from far away.
The earth looks like a big, round ball.

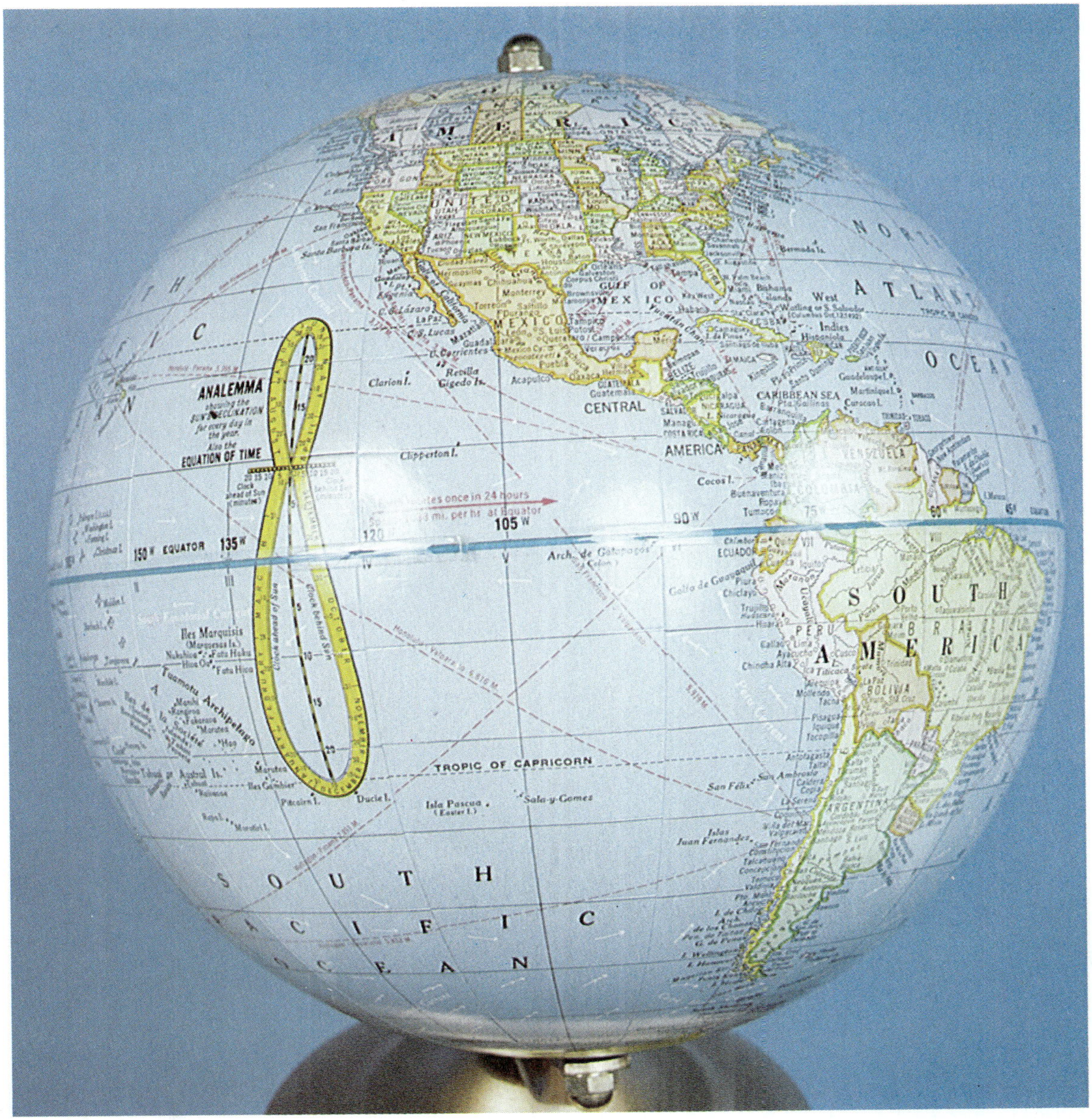

Here is something that looks like the earth.
It is like a ball, too.
How much can you see at once?

ACTIVITY

Which children are in the light?
This is like the day for them.
Which children are away from the light?
This is like the night for them.

Does the light always shine?

What happens as the children turn?

How can they go from night to day?

ACTIVITY

Find out where you live.

Put a small flag there.

Look at the side with the flag.

Is it in the light or dark?

Is it day or night?

Now turn it like this.
Watch the flag.
Point to the part that has day.
Point to the part that has night.

Now turn it all the way around.
Watch the flag.
Did you make one day and one night?

ACTIVITY

Sit on a seat that can turn.
What do you feel when you turn fast?
What do you see?

Now make a very, very slow turn.
What if you took one day to do it?
What would you feel?
What would you see?
The earth takes one day to turn once.

The earth turns.
It turns from night into day.
The earth is very big.
It makes only one turn each day.
You can not see it turn.
You can not feel it turn.

The sun always shines.
The earth keeps turning.
One side turns from dark to light.
The other side turns from light to dark.
The turning brings day.
Then it brings night again.

On some nights you can see the moon.
Where is the moon in the sky?
First it is here.
Later the same night it is there!
The earth turns.
The moon rises and sets.

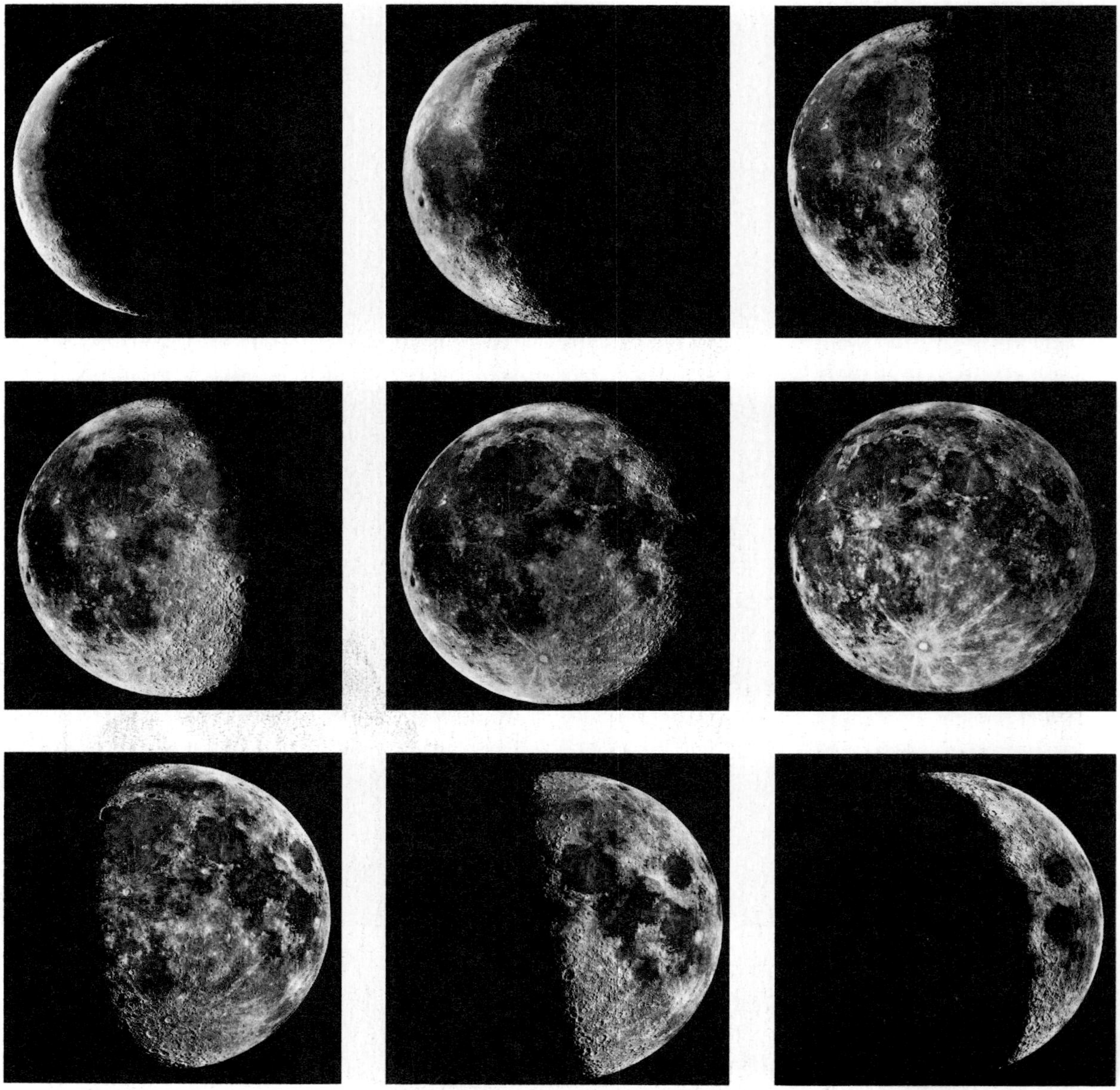

Have you seen the moon on different nights?
Did it look like any of these pictures?
The moon seems to change shape.

But the moon is round like a ball.
Where is the part you do not see?

ACTIVITY

Hold a ball in the light like this.

What is the shape of the lighted part?

Now sit like this.

Does the shape of the lighted part seem to change?

How is this like the moon?

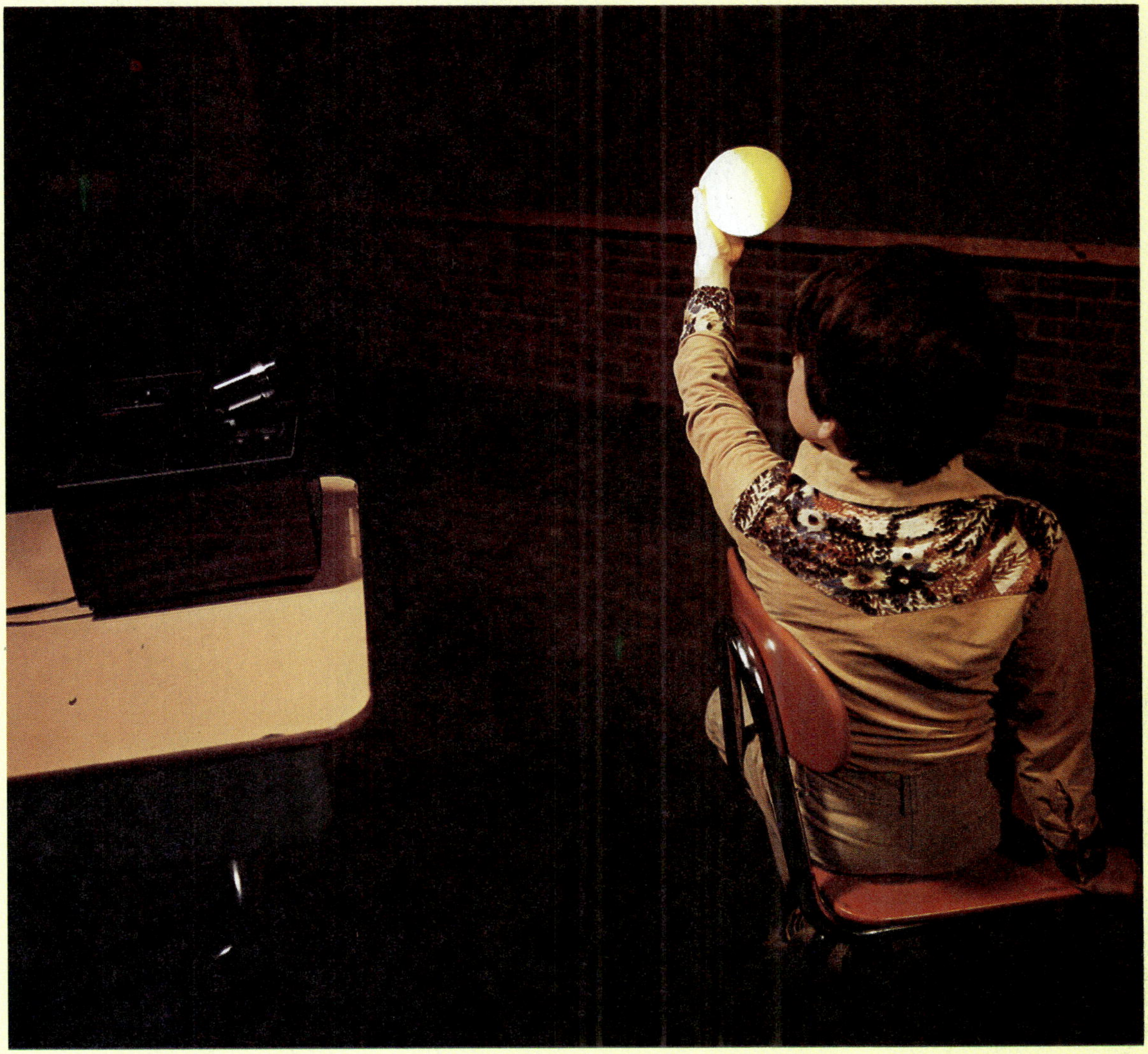

Some things stay the same.
The sun keeps shining.
The earth keeps turning.
Some things change.
Night changes into day.

Day changes into night.
The moon seems to change shape.
But even changes can be the same.
These changes happen again and again.

TO THINK ABOUT AND DO

FIND THE WORD

What word do you need?

You can use a word more than once.

sun	day
moon	night
earth	

The __?__ warms us and gives us light.

The __?__ gives us some light, too.

Day and night happen because the __?__ turns.

__?__ begins when the sun sets.

The __?__ seems to change shape.

CAN YOU TELL TIME BY THE SUN?

Look at the pictures.

This is where the sun is in the morning.

Where is it at lunch?

Where is it in late afternoon?

WHERE IS THE SUN?

Which child is right?

Can you tell why the child is right?

LIGHT AND DARK

This house is in the light.

The flag shows where the house is.

Which side of the earth is it on, light or dark?

Now the house is in the dark.

Now which side of the earth is it on?

7

Hot and Cold

These people are having breakfast.
Some things are hot.
Some things are cold.
How do we keep them hot?
How do we keep them cold?
How can we change them?

You can feel heat.
Heat is a kind of energy.
Energy can change things.
We get heat energy from the sun.

Where else do we get heat energy?

The sun heats some homes.
What heats your home?
No matter what you use, you are using energy.

ACTIVITY

Feel the cup.

Is it hot or cold?

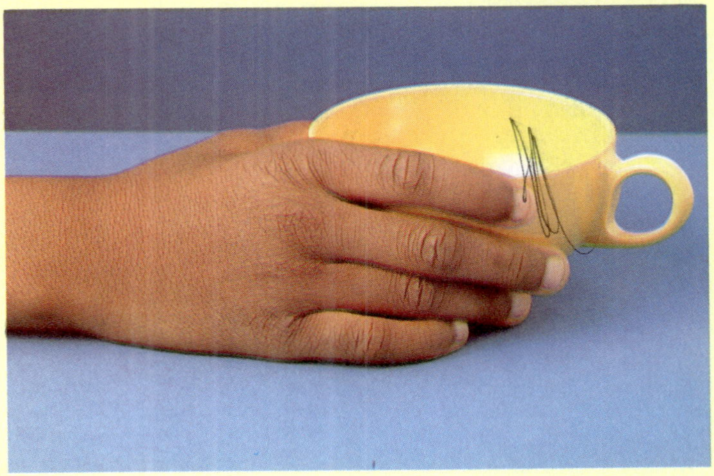

Now put warm water into it.

Feel the cup again.

What is happening to the heat?

Feel the air over the cup.

What did you find out?

Heat can move from one thing to another.
But sometimes we do not want heat to move.
Sometimes we want to keep the heat inside.
Other times we want to keep the heat outside.
We can do things to keep heat from moving.

Have you ever used one of these?
It keeps hot things hot.
It keeps cold things cold.

You give off heat.
What do your clothes do for you?
Which clothes are better for cold weather?

ACTIVITY

What is the best way to keep things cold?

Let's find out.

Try to keep the ice from changing.

Which ice changed first?

Which ice changed last?

How can you tell how hot something is?
One way to tell is to feel it.

ACTIVITY

Put one hand in the cold water.

Put the other hand in the warm water.

Keep your hands in the water for a few minutes.

Now put both hands in the other bowl.

What do you feel?

Sometimes feeling is not a good way to find out.
Why do we need a better way?

Where have you seen these things?
They show you how warm something is.
They are called thermometers.

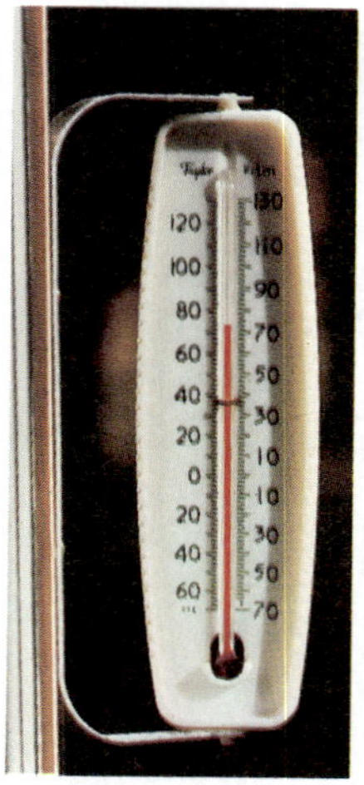

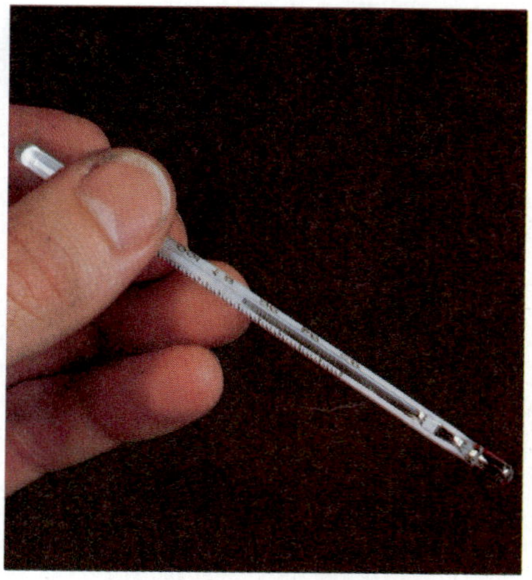

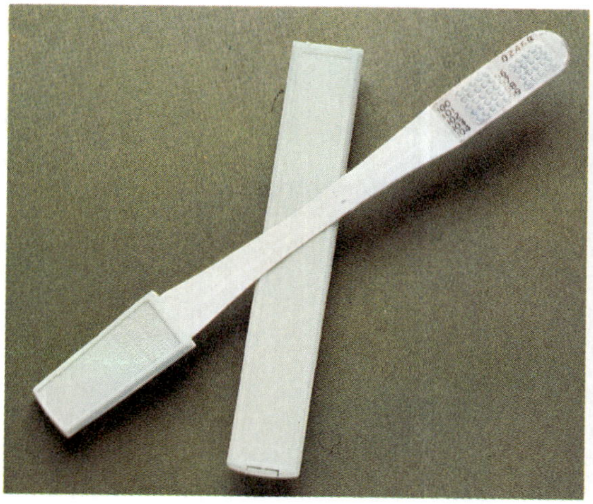

Some thermometers are used in the kitchen.
Other thermometers are used when you are sick.
What other kinds have you seen?

ACTIVITY

Look at a thermometer.
Find the color inside.
Where does the color stop?
It shows you how warm something is.
Air is all around the thermometer.
How warm is the air?

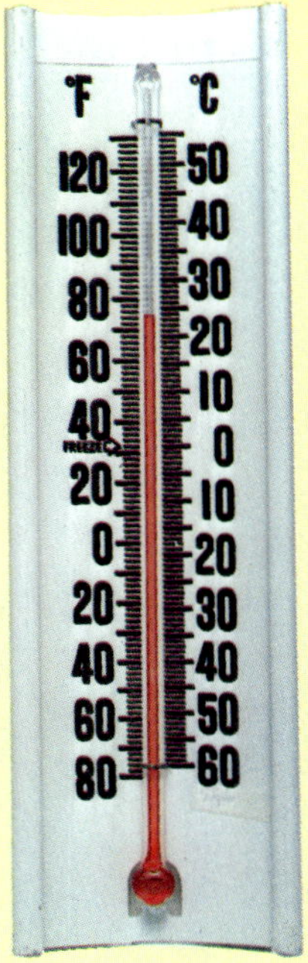

Put the thermometer in the ice.
Where does the color stop?
Now you can tell how warm or cold something is.

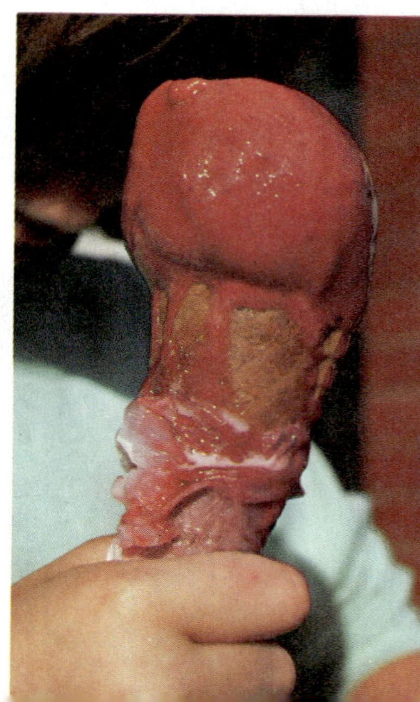

We use heat energy to change things.
What is changing in each picture?
Why do we want to change food?
Which changes are not wanted?

How do we keep food from changing?

People use heat and cold in many ways.
What hot or cold things have you used today?

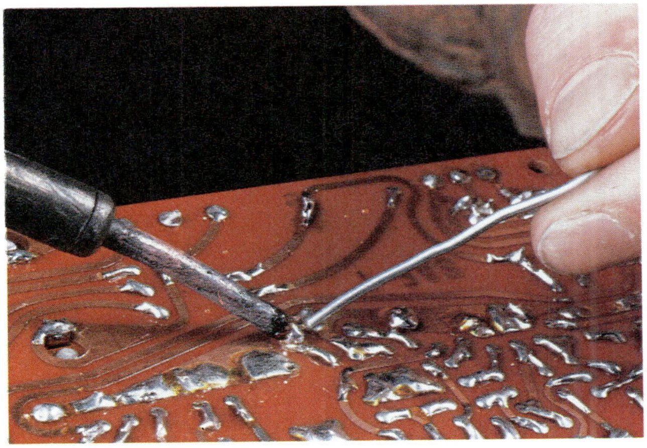

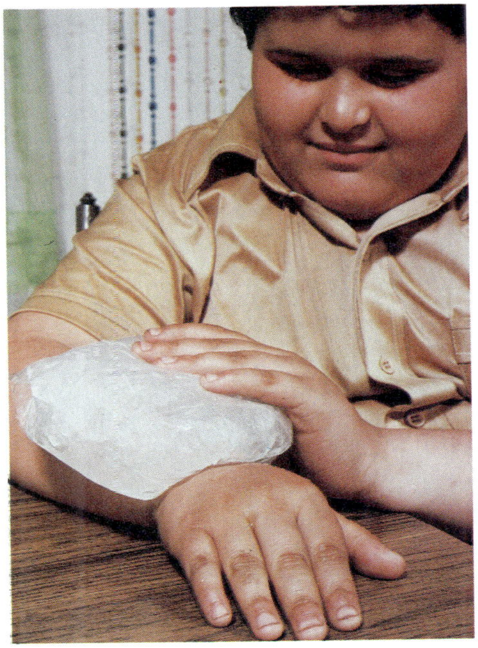

TO THINK ABOUT AND DO

THERMOMETERS

Look at the thermometers.

These words will help you.

hot   cold   warm

A shows that it is __?__.

B shows that it is __?__.

C shows that it is __?__.

Now look again.

Can you read where the red lines stop?

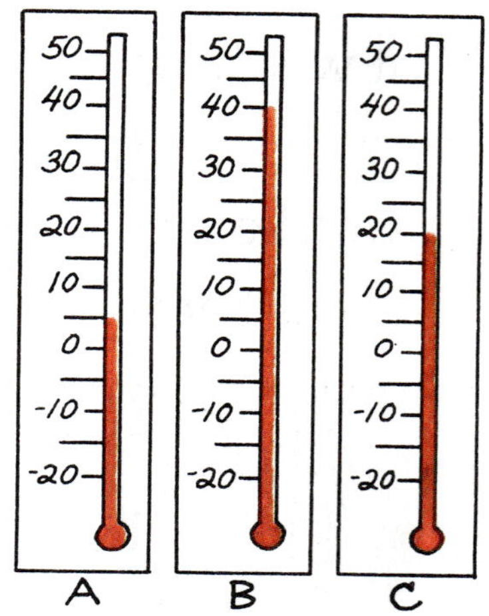

HOW DOES HEAT CHANGE THINGS?

Which pictures belong together?

HOT PLACES AND COLD PLACES

This girl has a thermometer at place 1.

How will it change if she moves it to place 2?

What will happen when she makes these moves?

2 to 8   8 to 6   1 to 8   3 to 1   7 to 4

Now make up some moves of your own.

Tell what happens.

8

Magnets

Look what happened!
How are we going to clean this up?
Can a magnet help us?
Will it pick up the nails?
Will it pick up the wood, too?
Let's find out.

Magnets have many shapes.
Some are long and straight.
Others are round and thin.
They have many shapes for many uses.
These toys use magnets.

ACTIVITY

Let's have fun with magnets.

Walk around the room with a magnet.

Touch some objects with it.

What did you find out?

What kinds of objects does your magnet pull?

Put some of these on a table.

Lay your magnet on top of them.

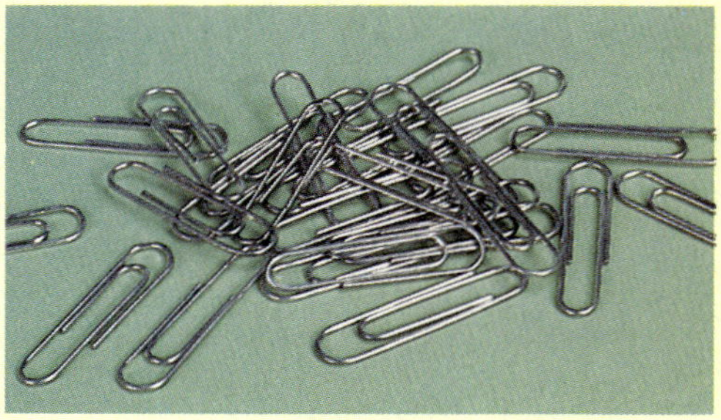

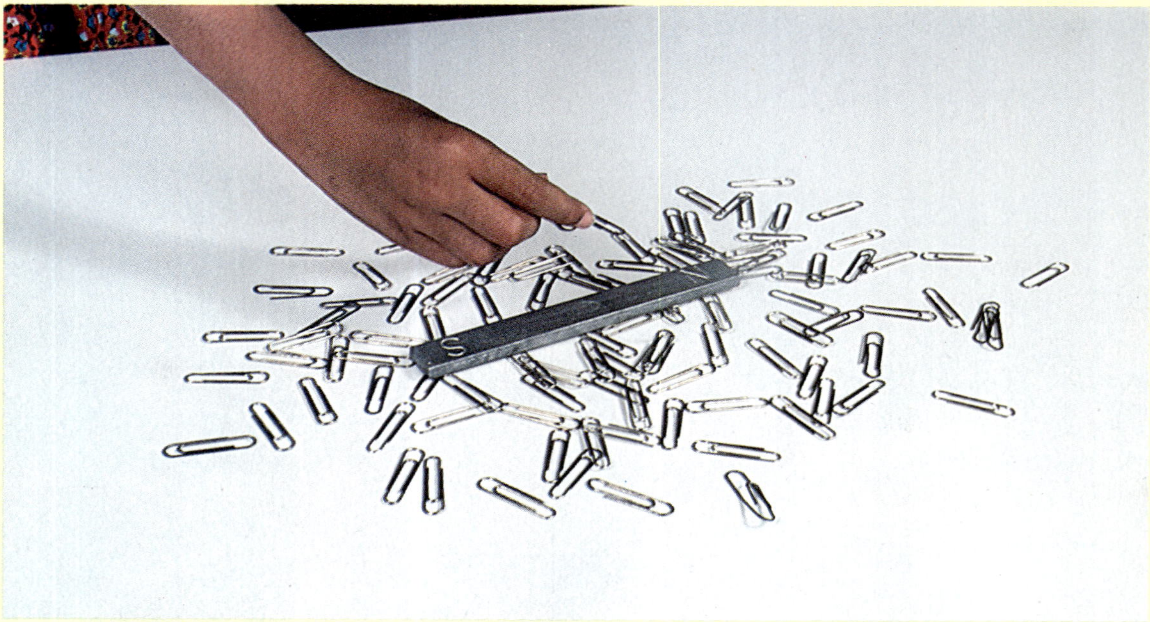

Now pick up the magnet.

What part of your magnet holds the most?

What part holds very few?

Try some other magnets.

What did you find out about magnets?

ACTIVITY

Try it again.
But now put paper over them.
Will the magnet pull through the paper?

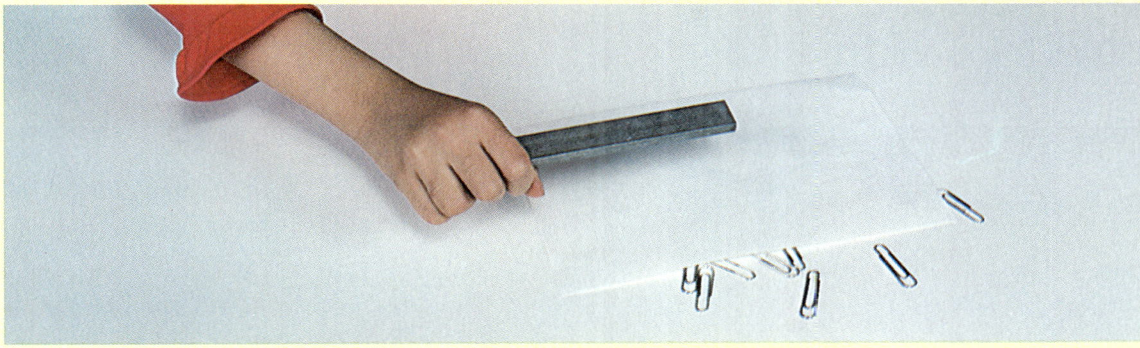

Here are some other things.
See if your magnet can pull through them.

ACTIVITY

Now try this with another child.

Put your two magnets close to each other.

What do they do?

Turn one around.

What happens?

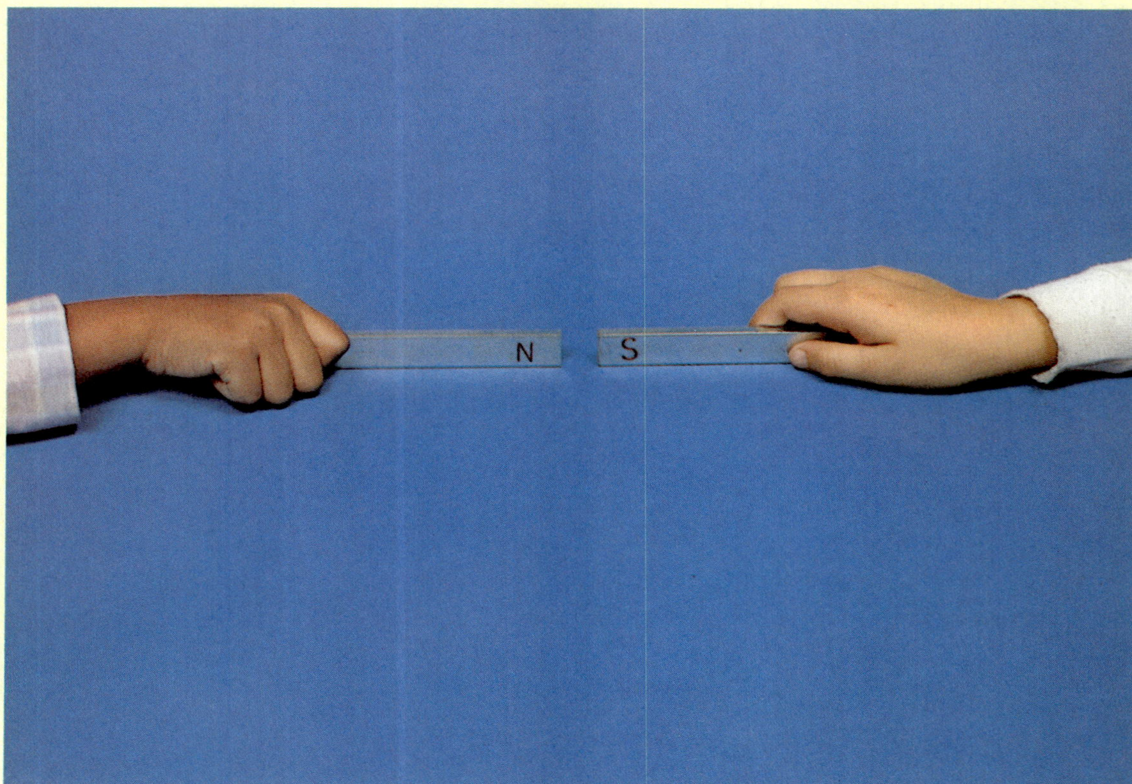

When do the magnets pull each other?

When do they push away?

Four children and only one magnet!
Each one wants to use it.
Can we make more magnets?

ACTIVITY

Here is a way to make a magnet.
Put the needle on the magnet.
Rub the needle along the magnet to the end.
Do this about ten times.

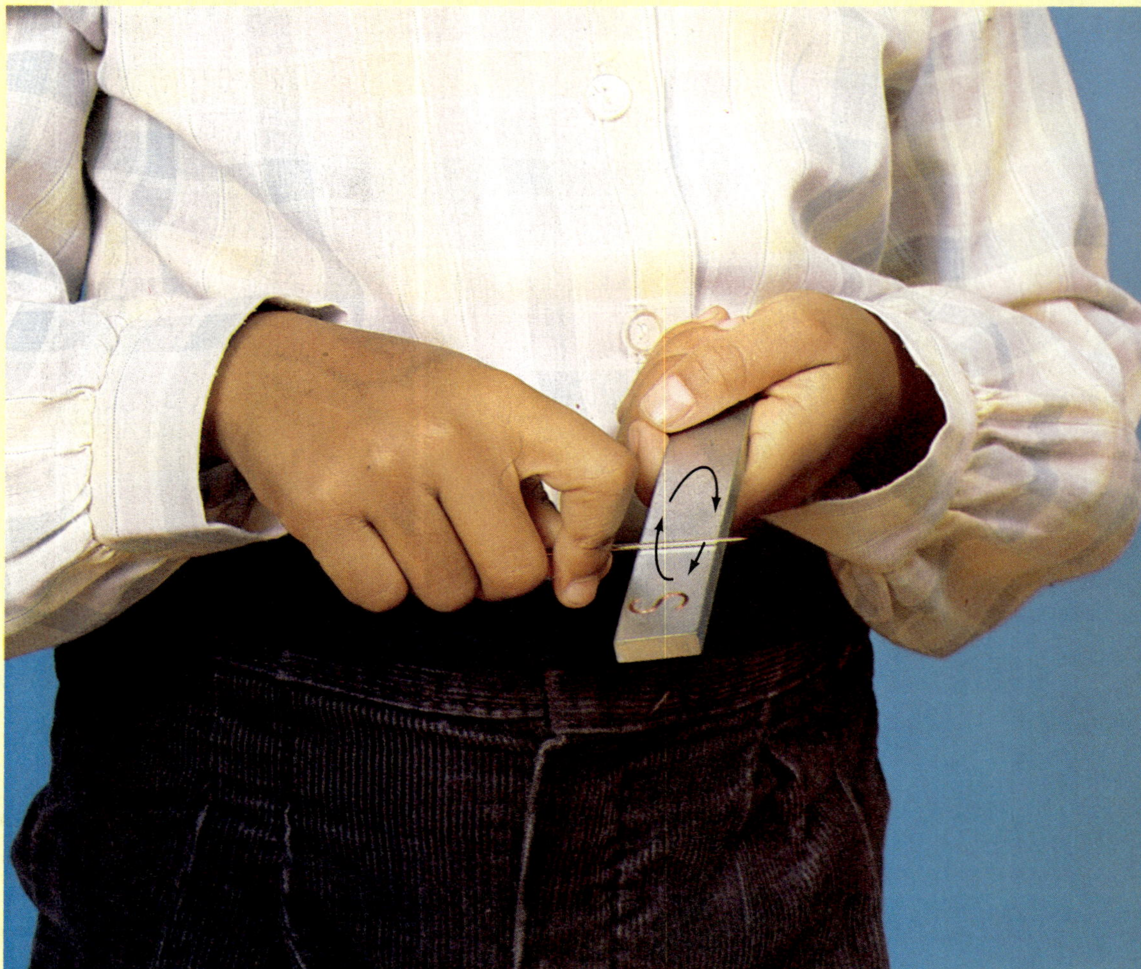

Have you made a new magnet?
How can you tell?
Try it!
Will it pull the same things?
Will it pull as hard as the first magnet?

ACTIVITY

Put your new magnet on a small piece of wood.

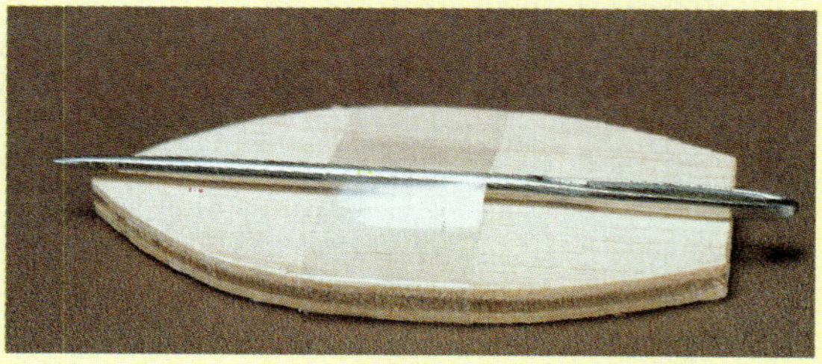

Place it in water.
Keep it away from the sides.
Let it turn until it stops.

Make a line on the side where the magnet points.
Then turn the magnet away.
Let it turn and stop again.
Make another line.
Try it again and again.
Does the magnet always point the same way?

Here is something else that always points the same way.

It is a compass.

It has a magnet, too.

One end always points north, as your magnet does.

These people are not lost.
They have a compass.
They can find out where north is.
Then they will know east and west, too.
They know which way to go.

There are no streets or road signs here.
Yet they know which way to go.
They have a compass.
They can find their way.

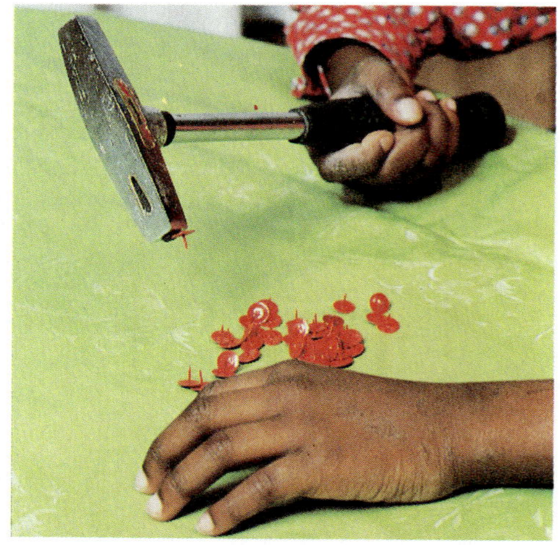

We use magnets in many ways.
Try this.
Go on a magnet hunt!
See how many other magnets you can find.
See how many ways we use magnets.

"MAGNETIC PERSONALITY"
CHANGE HIS CHARACTER WITH THE WAND

Jason
Smit

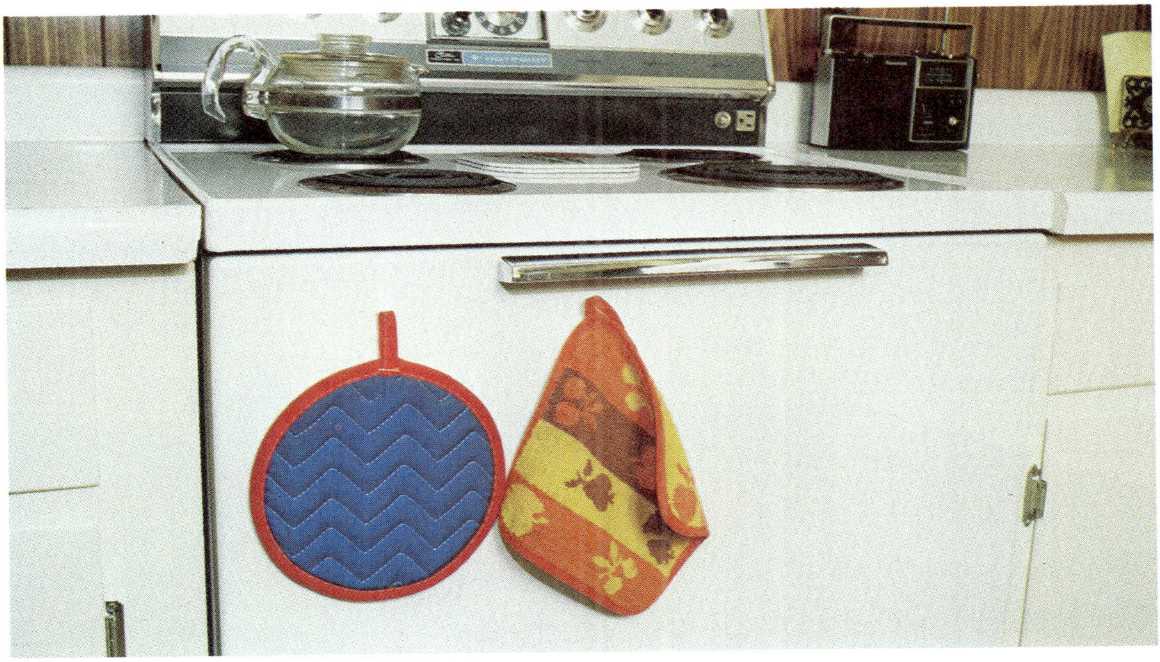

TO THINK ABOUT AND DO

WHAT HOLDS IT UP?

Why does it not fall?

Try these things.

CAN YOU DO THIS?

Put these in a cup of water.

Can you get them out without getting wet?

HOW STRONG IS YOUR MAGNET?

How many of these can a magnet pick up?
How many can stay in one line?
Try it!

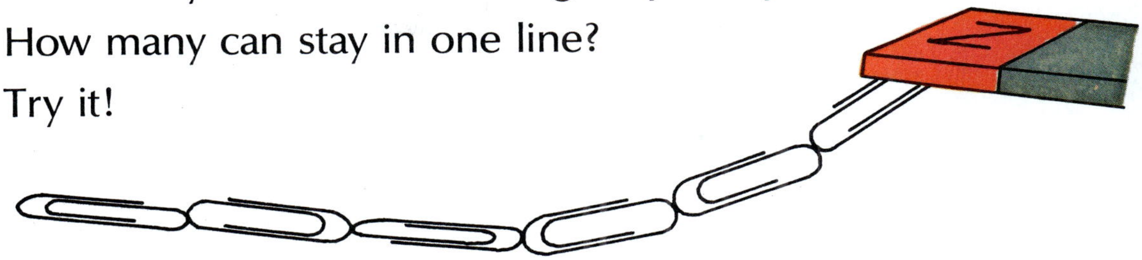

MAKE A BOAT THAT MOVES

Rub this with a magnet. 
Put the new magnet onto your boat.
Put it in water.
Use a magnet to move your boat.
Make it move to you.
Make it move away from you.
Have a race with other boats.

CAN YOU FIND THESE WORDS?

Move the letters around.
magnet, compass, push, pull, north

hups
lupl
tronh
tagnem
pascoms

9

Plants and Animals on Land

Many plants grow in this hot, wet place.
You can see more green leaves than you can count.
You can not see any animals.
But you can hear them.
You can hear many different sounds.

There are many kinds of plants and animals.
And they live in many places.
Some places are hot.

Some places are cold.

Some places are dry.

Some places are wet.

Some plants and animals live in the city.
What kinds can you find there?

The desert is a hot, dry place.
There are many days without a drop of rain.
How can anything live in such a place?

Many things do live here.
Look at the desert at night.
How many plants do you see?
How many animals do you see?

Here is the same place in the day.
The sun is very hot and bright.
Where did all of the animals go?

Roadrunner

Chuckwalla

In the day some animals stay in the shadows.
Many animals stay in holes in the ground.

Kit foxes

Then night comes again.
It is not as hot outside.
Many animals come out again.

ACTIVITY

Dig a deep hole in some soil.
Put one thermometer in the hole.
Cover just the end of the other thermometer.
How warm is each place?

Now put the pot in the sun.
Wait a while.
Now how warm is each place?
Where would animals go on a hot day?

House finch

Cactus wren

Mourning dove

Gila woodpecker

Leaf-cutting ant

Tarantula

Cactus joint bug

Scorpion

Look at these desert animals.
What do you know about them?
Why can these animals live in the desert?

Desert tortoise

Side-blotched lizards

Sidewinder rattlesnake

Horned-toad lizard

Coati

Chipmunk

Bighorn sheep

Bobcat

Desert soil has very little water.
Desert plants have to keep water inside.
What shapes do these plants have?
Which plants can live in the desert?

ACTIVITY

Find two pieces that are alike.

Wet them.
Make one into a ball.
Put them up like this.
What do you think will happen?

Feel both pieces the next day.
Which one is still a little wet?
Why can a plant shaped like this live in the desert?

ACTIVITY

You can make a small desert.
What plants would you put in it?
What animals would you put in it?

The woods have many tall trees.
Many plants and animals live in the shadow of the trees.
It can be hot and sunny in other places.
But the woods stay quiet and dark.

There are rocks, logs, and leaves on the ground.
What does this do to the soil?

ACTIVITY

Put soil in two pans.
Wet the soil in both pans.
Cover the soil in one pan with leaves.
Put the pans in the sun.
Wait a few days.
What happens?

Moss, fern, and sorrel

Shelf fungus

Moss

Lady slipper

Ground pine

There are many small plants in the woods.
How many of these have you seen?
How are they different from the trees?
They do not need much light.
And they need wet soil.
Many plants could not live here, but these can.

Mushroom

Some animals live in the wet soil.
Look under a rock, a branch, or a log.
What animals do you see?
What do they look like?

Many other animals live in the woods.
Some live in the trees.
Some live on the ground.

Some even live in the ground!
They eat, sleep, and grow here.
Tell about each animal.

ACTIVITY

You can make your own small woods.

What plants would you put in it?

What animals would you put in it?

There are many kinds of plants and animals.
They live in many places.
But they can not all live in the same place.
Some can live best in a desert.
Some can live best in a woods.
There are different ways to live in different places.

TO THINK ABOUT AND DO

CAN YOU SPELL THESE WORDS?

What are the missing letters?

pl _?_ _?_ ts

an _?_ _?_ als

d _?_ _?_ ert

w _?_ _?_ ds

gr _?_ _?_ nd

THE WOODS AND THE DESERT

Where do these animals live?

Put the animal with the picture of its home.

Can any of these animals live in both places?

PLANT A TREE

Have you seen seeds like these?

Find some seeds.

Plant the seeds like this.

What will they need to grow?

Watch them grow.

LOOK AT THESE PLANTS

Which ones live in the desert?

Which ones can be found in the woods?

How are the plants different?

10 Plants and Animals in the Water

Have you ever visited a pond?
The water is still.
The air is quiet.
You can hear a buzz.
Something is flying by.
Splash!
A shadow moves under the water.
Many things are happening here.

A green frog is sitting on a green plant.
Can you see it?
It is hard to see a frog when it sits still.
How does the frog's color help it?

A frog can swim in the water.
It can also hop on land.

Green frog

Spotted turtle

Wood duck

Muskrat

Great blue heron

A pond is a land and water place.
Many pond animals live in both places.
Which of these animals can live in both places?

Long-tailed salamander

Brown bullhead

River otter

Water snake

Moose

Dragonfly

Waterbug eating damselfly

Yellow jacket

Diving beetle

There is a lot of food at the pond.
Look at all of these animals.
Some of them eat plants.
Some of them eat animals.

Water strider

Water tiger eating tadpole

Dragonfly

Water scorpion

Bog fritillary butterfly

Caterpillar

Horsefly eggs

Mosquito eggs

Fly laying eggs

Many of them lay their eggs here, too.
Why are there so many of them?

Water striders

Mayflies

Mallard ducks

These ducks are busy looking for food.
They are digging in the mud.
The duck's bill keeps the food in.
It also lets the water and mud come out.

Black duck

Common merganser

Redheaded woodpecker

This bird's bill has something like teeth.
Can a fish slide out of the bill?

Look at all these other birds.
Tell how they use their bills.

Atlantic puffin

Roseate spoonbill

Anhinga

Great white heron

Osprey

How does this bird use its bill?
How does it use its claws?

Mallard

Ring-billed gull

Look at the shape of these feet.
Why are they good for swimming?
Why are they good for walking on mud?

American avocet

Look at the long legs of these birds.
Why are they good for walking in the water?

Beaver

Here is another animal.
How does it use its feet and tail?
How does it use its teeth?

A small river used to be here.
These animals put trees and mud across it.
The trees and mud made the water stay back.
This helped to make the pond.

What other animals live near the pond?
How are they alike?
How are they different?

ACTIVITY

You can make a pond.
What animals will you need?
What plants will you need?

These animals live in ponds.
Look at one.
How does it move?
What does it eat?
Does it have eyes?
What happens if you tap on its shell?

Many ponds have fish.
Look at some fish.
What do fish eat?

What parts help them move?
Watch your pond every day.

Have you ever been here?
How is the water different from pond water?

ACTIVITY

Put some of this and water in a cup.
Put only water in the other cup.

Now taste a little water from both cups.
How is the water different?
Now get two more cups.
Do this.

Let the water on each cup dry.
What do you see on each cup?

Many plants and animals live in the water.

Starfish and sea anemones

Sea urchin

Barnacles and periwinkles

Starfish

Seaweed

Some of the plants and animals are small.
Have you seen any like these?

Some animals are big like these.

How big are they?

How can you tell?

Elkhorn coral

Florida sponge

Green sea anemones

Acorn barnacles

Some animals stay in one place.

Coquina clams

Octopus

Coral shrimp

Hermit crab

Atlantic lobster

Nine-armed sea star

Some animals move along the bottom.

French grunts

Black-barred soldierfish

Yellowtail damselfish

Spotted puffer fish

Swellfish

Snappers

Some animals move a lot.
Many fish of every color swim by.
The water is full of living things!

Some of these animals eat plants.
Some of them eat animals.
Tell what is happening in the picture.

We get food from the water.
Which of these foods do you like?

So many different animals and plants!

Yet each has found ways to live in or near water.

TO THINK ABOUT AND DO

ANIMAL PARTS

Water animals have many parts that help them.
Move the letters around to find the names of these parts.
The pictures will help you.
Look at the lines around the parts.

hetet
illb
slwac
lita
slehl

POND ANIMALS

Make a list of animals that live at a pond.
Find out what they eat.
How do their bodies help them move?

Pond Animals

Name	What They Eat	How They Move

NAME THE POND ANIMAL

I can swim in water.

I can hop on land, too.

I can be hard to find during the day.

You can hear me at night.

Draw a picture of me.

SOMETHING DOES NOT BELONG

In each picture one animal does not belong.

Can you find it?

11 Plants and Animals of Long Ago

No one has ever seen one of these animals.
They lived long, long ago.
There were many of them at that time.
Do you know what they are called?

Dinosaur bones

They are dinosaurs!

People have found parts of dinosaurs.

Tyrannosaurus Triceratops Trachodons

These parts can be put together like this.
This tells us something about the dinosaur's shape.

Model of tyrannosaurus rex

This also tells us about how big they were.
So now we think we know how they looked.

Brontosaurus

There were many kinds of dinosaurs.
Some dinosaurs were very big.
Some were as tall as a house.

Diplodocus

Others were as long as two buses.

Brachiosaurus

Some big dinosaurs were as heavy as all of these.

Turkey Heterodontosaurus

Others were about as small as a large bird.

Dinosaur footprints

Here is where some dinosaurs walked through mud.
Over many years the mud turned into rock.
The rock shows the shape of their feet.

Pteranodon Pleisiosaur

Bowfin Garfish

How do you think they moved?
Some animals could walk.
Some could swim.
Others could fly.

Plant fossil

Plant fossil

Plant fossils

Petrified wood

The rocks show plants that lived at the same time.
We know that the world looked very different then.

Anatosaurus Tyrannosaurus

Some dinosaurs ate the plants.
Other dinosaurs ate animals.
There are no dinosaurs living today.
We can not be sure why this happened.
This may happen to other plants and animals today.

Wooly mammoth

Eohippus

Saber-toothed tiger

There are no more of these animals living today.

Steller's sea cow

Dodo bird

Great auk

How might this have happened?

This bird was the last one of its kind.
It was in a zoo to save it.
Now there are no more like it.

Martha, the last passenger pigeon

Pink lady slippers

Prairie white fringed orchids

Mead's milkweed

Globe mallow

Water plantain

French's shooting star

Franklinia

Turk's cap lily

Only a few of these plants are left.

Timber wolf

Grizzly bear

Ocelot

Hawaiian ne-ne geese

Peregrine falcon

Bald eagle

And only a few of these animals are left.

Brown pelican

Whooping crane

California condor

San Francisco garter snake

American alligator

Bison herd

Many changes happen where plants and animals live.

Right whales

Tropical rain forest

What changes do you see in these pictures?

At one time there were many kinds of dinosaurs.
Now there are none.
Many plants of long ago are gone, too.

This might happen to some plants and animals you know.
What can we do to save them?

TO THINK ABOUT AND DO

DINOSAURS WERE HERE!

You can see where they walked.
You can make something like it.
Put some of this on your hand.
Then do this.
Let it get hard.

DINOSAUR HUNT

Pick the dinosaur you like best.
Find out more about the dinosaur.
What did you learn?

WHAT DID DINOSAURS LOOK LIKE?

Find some clay.

Make a dinosaur.

Now think about where dinosaurs lived.

Make a place for your dinosaur.

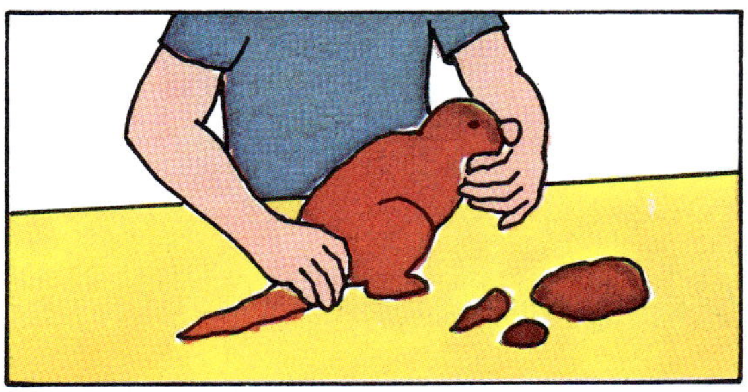

LOOK ALIKES

Here are animals of today and long ago.

Which ones lived long ago?

Which ones are living today?

SCIENCE WORDS

CREDITS

Illustrations Laurel Aiello: 20, 21, 38, 39, 60, 61, 62, 63, 83, 84, 85, 102, 103, 124, 125, 144, 145, 162, 163, 190, 191, 222, 223, 246, 247 / John Hamberger: 7, 12, 172, 186-187, 189, 206-207, 218, 220, 221, 224, 229, 230, 231, 233, 235, 236, 237 / James Teason: 31, 51

Photography J. C. Allen and Son, Inc.: 13 lower right, 17 lower right / A.W. Ambler, Photo Researchers: 238 lower middle / The American Museum of Natural History: 227 / Erik Anderson: 53, 57, 89, 95 / R. Austing, Bruce Coleman, Inc.: 200 lower right / Ken Balcomb: 243 upper right / Jen and Des Bartlett, Bruce Coleman, Inc.: 243 upper left / The Bettmann Archive, Inc.: 238 bottom / S.C. Bisserot, Bruce Coleman, Inc.: 200 lower left / Fredrik Bodin: 8 top, 143 lower left / A. James Casner: 160 lower left / Cincinnati Zoo: 238 top / Stuart Cohen: 70 lower right / Michael Collier: vi upper left, lower left and right, 35 bottom, 88 lower left, 232 lower left, 234 / Dr. E.R. Degginger: 115, 171, 174 bottom, 197 upper right, 198 lower left, 199 upper right, 202 lower left, 213 lower right, 215 lower left, 216 upper right, middle and lower left, 239 upper left, lower right, 240 middle left, upper right, 241 upper left / Marvin L. Dembinsky, Jr.: 199 lower right, 239 upper middle / Peter H. Dreyer: 24 upper left / Tom Duerst: 17 lower left, 198 middle right, 199 middle and lower left / Silvia A. Earle, Seafilms, Inc.: 214 / Harry Engels, NAS, Photo Researchers: 204 upper right / Nancy Fenton: 8 bottom / William E. Ferguson, Nature Photography: 174 upper left and right, 176 middle right, lower middle and right, 177 upper left and right, middle left and right, 200 middle left / Field Museum of Natural History, Chicago: 232 right / Animals Animals © Henry Fox: 201 bottom / Freelance Photographers Guild: 22 / © Rick Friedman, The Picture Cube, 142 bottom / © Bill Gallery, Stock, Boston: 70 upper left / Howard Goldbaum: 217 lower right / Luther C. Goldman, Bureau of Sport Fishing and Wildlife, U.S. Department of the Interior: 241 upper right / Patrick Grace, Photo Researchers: 203 lower left / Grant Heilman: 129 upper left, 167 / Anne Heimann: 17 upper right, 40, 195 top, 197 middle right / Barry Howe: 164, 169, 181 / Harold R. Hungerford and Associates: 173, 176 top left and right, middle left and middle, 177 lower left, lower middle and right, bottom left, 200 upper right, 215 middle left, 216 upper left, middle and lower right, 217 lower left, 239 upper right, middle and middle right, lower left / Ray Hunold: 93 upper right / Imagery: 25 middle left, 88 upper right / Paul E. Johnson: vi upper right, 2, 68 upper right, 82 upper and middle right, 86, 100, 101 upper right, 129 lower left, 199 upper left, 213 upper right, middle left, 219 bottom / Lou Jones: 137 lower right, 140 lower left / Kendal Whaling Museum: 157 upper right / Breck Kent: 3 left, 107, 195 bottom, 196 upper left, 197 upper left, 213 upper left, 226, 240 lower left / A. Kerstitch, Taurus: 217 upper and middle left / Stephon J. Krasemann, Photo Researchers: 203 lower right / Stephen W. Kress, NAS, Photo Researchers: 203 top / © Susan Lapides: 219 bottom / Calvin Larsen, Photo Researchers: 202 lower middle / Harriet Lee: 96 top / Zig Leszczynski: 196 lower right, 198 lower right / Lick Observatory: 118, 119 / Robert Llewellyn: 82 lower right / Talbot Lovering: 78-79, 140 middle left / Thomas Magno: 131,

139, 152, 154 / Michael Malyszko: 228 / Anthony Mercieca, Photo Researchers: 202 upper left / Dan McCoy, Rainbow: 25 lower left, 93 upper left, 176 lower left, 204 bottom / © Bruce McDonald, Animals, Animals: 204 upper left / Tom McHugh, Photo Researchers: 241 middle left and right / Ralph Mercer: 5, 9, 14, 15, 26, 27, 28, 29, 33, 42 lower left, 45, 47, 48, 50, 54, 55, 56, 68 upper left, 70 upper right, 72, 73, top, 76, 77 bottom, 96 bottom, 109, 129 lower right, 133, 134, 135 upper right, 136, 137 upper and lower left, upper right, 138, 140 upper right, middle right, lower middle, 146, 148, 150, 151, 152, 155, 156, 157 upper left, bottom, 180, 183, 185, 188, 208, 209, 211 / Hank Morgan, Rainbow: 24 lower left, 68 lower right, 93 bottom / NASA: 108, 116 / Palmer-Brilliant: 140 upper left, lower right, 142 middle right, 143 upper left, lower right / Thomas Pantages: 3 lower right, 11, 13, upper right, 82 lower middle, 142 upper right, 198 top / Norman Prince: 25 middle and lower right, 184 upper and middle right, 213 lower left / Lawrence Pringle: 184 lower middle / Herbert Randle: 101 bottom / © Eric A. Roth, The Picture Cube: 24 lower right / Leonard Lee Rue III: 13 lower left, 166, 196 upper right, lower left, 197 middle left, bottom, 201 top, 202 middle right, 205 lower left, 215 lower right, 240 upper left, middle and lower right, 241 bottom, 242 / L. L. Rue III, Photo Researchers: 205 upper left / John Running: 137 middle right, 219 top / Gil Sappah: 215 top / Peter Schweitzer: 58 top / Frank Siteman: 6, 10, 35 top, 132 left, 170 / Slidemakers: 25 upper right, 68 lower left, middle right, 70 lower left, middle right, 129 upper right, 142 upper and middle left, 143 middle left, 232 upper left / Camilla Smith, Rainbow: 70 lower middle, 168 / Diedra Delano Stead: 32, 104 / Larry Stepanowicz: 199 middle right / John Swedberg: 205 upper, middle, lower left / Margaret K. Tenney: 58 bottom, 123, 130, 132 right, 184 upper left / William Tenney: 4, 13 upper left, 17 upper left, 24 upper right, 88 lower right, 101 lower left, middle, 122, 129 middle left, 167, 182, 184 lower left and right, 192, 243 lower left and right / Titmus, Taurus: 88 upper left / Bonnie Unsworth: 88 lower middle / Unsworth-O'Neil: 16, 18, 30, 34, 36, 37, 42 top and lower right, 43, 44, 46, 49, 52, 64, 66, 67, 69, 71, 73 bottom, 74, 75, 77 top, 80, 81, 82 upper and lower left, 90, 91, 92, 94, 97, 98, 99, 106, 110, 111, 112, 113, 114, 117, 120, 121, 126, 127, 135 upper right, lower left and right, 141, 149, 153, 158, 159, 160 upper left and right, lower right, 161, 175, 178, 179, 210, 212, 244 / R. Terry Walker: 143 upper right / Wallin, Taurus: 217 middle right / Animals Animals © Fred Whitehead: 202 lower right / Muriel V. Williams: 3 upper right, 200 upper left / Ron Willocks, NAS, Photo Researchers: 202 upper right / Ellen Woodberry: 88 middle right, 194 / Dave Woodward, Taurus: 217 upper right / Barbara Young, Photo Researchers: 59

Special thanks are extended to the following schools and organizations which were helpful in the production of photographs: Muraco Elementary School, Winchester, Massachusetts; Lexington Public Schools, Lexington, Massachusetts; Oak Hill Elementary School, Newton, Massachusetts; Boston Museum of Science, Boston, Massachusetts, North Country Outfitters, Concord, Massachusetts

1 2 3 4 5 6 7 8 9 0